1 MONTH OF
FREE
READING

at
www.ForgottenBooks.com

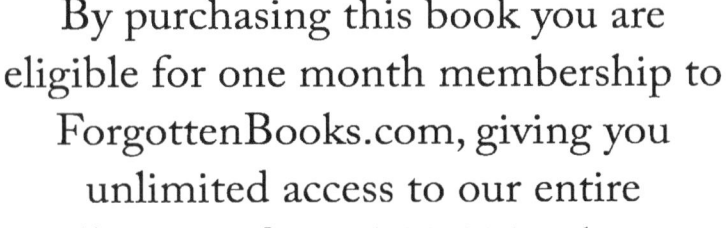

By purchasing this book you are eligible for one month membership to ForgottenBooks.com, giving you unlimited access to our entire collection of over 700,000 titles via our web site and mobile apps.

To claim your free month visit: www.forgottenbooks.com/free309584

ISBN 978-0-483-16877-0
PIBN 10309584

This book is a reproduction of an important historical work. Forgotten Books uses
state-of-the-art technology to digitally reconstruct the work, preserving the original format
whilst repairing imperfections present in the aged copy. In rare cases, an imperfection in
the original, such as a blemish or missing page, may be replicated in our edition. We do,
however, repair the vast majority of imperfections successfully; any imperfections that
remain are intentionally left to preserve the state of such historical works.

ERNEST·SOLVAY
ORIGINATOR AND PERFECTOR OF THE
SOLVAY PROCESS
FOR THE MANUFACTURE OF CARBONATE OF SODA

ROWLAND HAZARD
FOUNDER AND FIRST PRESIDENT OF THE SOLVAY PROCESS CO.
SYRACUSE, N. Y., 1881

WILLIAM B. COGSWELL
ASSOCIATED FROM ITS ORIGIN WITH
THE SOLVAY PROCESS CO.
AS
ENGINEER, GENERAL MANAGER, TREASURER, DIRECTOR AND VICE-PRESIDENT

FREDERICK R. HAZARD
PRESIDENT OF THE SOLVAY PROCESS COMPANY

MAIN OFFICE OF THE SOLVAY PROCESS COMPANY, SYRACUSE, N. Y.

SOLVAY ALKALI

ITS VARIOUS FORMS & USES

WITH NOTES ON ALKALIMETRY AND CHEMICAL
AND COMMERCIAL TABLES CONVENIENTLY
ARRANGED FOR THE USE OF THE CONSUMER

1916

THE SOLVAY PROCESS COMPANY

SYRACUSE, N. Y.

THE SOLVAY PROCESS COMPANY

MANUFACTURER OF

SODA ASH, CAUSTIC SODA, BICARBONATE, CRYSTALS,

AND ALLIED PRODUCTS

OFFICERS:

PRESIDENT, F. R. HAZARD

VICE-PRESIDENT, R. G. HAZARD	GENERAL MANAGER, J. D. PENNOCK
VICE-PRESIDENT, W. B. COGSWELL	MGR., DETROIT WORKS, A. H. GREEN, JR.
VICE-PRESIDENT, E. N. TRUMP	CHIEF ENGINEER, C. G. HERBERT
SECRETARY, LOUIS KRUMBHAAR	ASS'T-TREASURER, R. W. SWIFT

WORKS:

SYRACUSE, N. Y.   DETROIT, MICH.   HUTCHINSON, KAN.

SELLING AGENTS:

WING & EVANS, INC., 22 WILLIAM STREET, NEW YORK

Organization

The Solvay Process Co. was incorporated in 1881 under the Laws of the State of New York, for the manufacture of soda in its various forms and for dealing in the raw materials and by-products of such manufacture. The Company has been in continuous operation since its incorporation. Its works are located at Solvay, near Syracuse, N. Y., at Detroit, Michigan, and at Hutchinson, Kansas.

The output of the original plant at Syracuse was 30 tons of soda ash a day. A study of the accompanying cuts may be found interesting as showing the tremendous growth of the industry. The Company today furnishes a very large part of all the alkali consumed in the United States.

From the origin of the business the sale of the Soda Products of the company has been under the direction of Wing & Evans, later succeeded by Wing & Evans, Inc., of 22 William Street, New York City, as General Selling Agents.

Wing & Evans, Inc., like the Solvay Process Co., has been continuously identified with the alkali industry of the United States from its beginning.

M82532

Brief Outline of the Soda Process

SODA ASH was first produced commercially by the LeBlanc process about the year 1800. That process was crude and expensive, the price of soda ash made by it being from $190 to $380 a ton. As early, therefore, as 1811, when the LeBlanc process was still in its infancy, experiments were commenced to discover some process more economical and efficient. These experiments were carried on almost continuously by some of the most noted scientists of Europe until the year 1861, when two Belgians, Messrs. Ernest and Alfred Solvay, brothers, originated the apparatus which enabled them to perfect the process for the manufacture of soda ash by the reaction of ammonium bi-carbonate on salt brine. This process was put into actual commercial operation at a plant at Couillet, Belgium, by the Messrs. Solvay in 1863, and the successful manufacture of soda ash by it has continued ever since. The process has long been known the world over as the Solvay process, and probably 95 per cent. of all the soda ash now manufactured is made by it.

The service to the world of the Messrs. Solvay in perfecting their process can never be appreciated. Entering as soda ash does, in one form or another, into practically all the products of industry, the tremendous economy effected in its manufacture by the Solvay process has been of enormous benefit to the world.

At its origin The Solvay Process Company acquired the American patents of the Messrs. Solvay, whose broad-minded policy provided for the establishment of a fraternity of independent organizations, one for each of the great countries whose consumption of alkali warranted a home supply. These national organizations are entirely separate and independent; each is under exclusive home control, but it is the privilege of each to share in the improvements of all the others by the interchange of technical and factory reports and by personal visits of their staff experts.

The business world can hardly offer a parallel to this Solvay fraternity, and thirty-five years have demonstrated its wonderful efficiency. By this alliance the users of the product of The Solvay Process Company derive the benefit, not only of the ablest experts in the manufacture of alkali in the United States, but in all the world.

The original patents have now expired. Companies have been formed both in Europe and in this country to use such portions of the Solvay process as they can readily follow. Such followers are a tribute to the excellence of the process.

Solvay' Products

The Solvay Process Company are manufacturers of the following products:

SODA ASH.—All grades and densities.

CAUSTIC SODA.—All grades of solid and ground.

BICARBONATE OF SODA.—Pure and commercial grades.

SESQUI-CARBONATE OF SODA.—Pure.

MODIFIED SODAS (So-called Neutral Sodas).—All grades.

CAUSTICIZED ASH.—All grades.

CROWN FILLER.—Pure hydrated Calcium Sulphate.

CALCIUM CHLORIDE.—All grades of solid, granulated and liquid.

SALT.—Commercial Grades.

LIMESTONE.—Pulverized and graded sizes.

For details of the various products and their uses see following pages. Any of the special pamphlets mentioned may be had on request.

Soda Ash—Na_2CO_3

Soda Ash is found in commerce in the following grades, all of which are made by The Solvay Process Co.:

58% Ordinary and 58% Dense.
48%   "   "   48% Special.

The 58% Ash is the highest grade of Soda Ash manufactured, and contains about 99% Na_2CO_3.

The distinction between 58% Ordinary (or Light) and 58% Dense is merely one of density, the Dense Ash weighing about twice as much as the Ordinary Ash per unit volume.

Chemically, they are identical, and perform the same functions, the Dense Ash being used where small bulk is desirable, e.g., in glass manufacture.

USES:

In the manufacture of glass, soap, paper, chemicals, drugs, paints, leather, enamel ware, cleansers. Soda Ash is also used in the textile industries, in dyeing operations, water softening, metallurgical operations, bottle and dish washing, refining of vegetable and mineral oils, metal working, prevention of timber mold.

The 48% Ordinary Ash and the 48% Special are reduced with NaCl and Na_2SO_4 respectively, and are used for special purposes.

PACKAGES: See page 42.
See special chapter on Dense Ash for Glass Manufacture, on page 20.

Caustic Soda—NaOH

Caustic Soda is manufactured in the following grades:

SOLID CAUSTIC.—76%, 74%; 70% Ordinary,
                70% Special; 60% Ordinary,
                60% Special.

GROUND CAUSTIC.—76%, 74%.

Caustic Soda is graded according to the percentage content of actual alkali (Na_2O) in it, 76% being the highest commercial grade.

The chief uses of Caustic Soda are in the manufacture of soap, paper, chemicals and drugs, paints, enamel ware, leather; used also in the textile industries, mercerizing of cotton, water softening, bottle washing, vegetable and mineral oil refining, metal working, and in the preparation of cleansers.

The Special Caustic Sodas contain certain amounts of Sodium Carbonate and Sodium Sulphate, and are of a softer nature than the ordinary Caustic.

Ground Caustic is ordinary solid caustic ground for putting up in small packages, for use in cleansing, in batteries, etc.

PACKAGES: See page 42.

For special chapter on the Use of Caustic Soda in Soap Making see page 22.

Bicarbonate of Soda—NaHCO₃

BICARBONATE OF SODA in the pure form is the well-known Baking Soda. It is used also in the manufacture of baking powders, both of these products being used over the civilized world.

Other grades, not so highly refined, are used for producing Carbonic Acid for charging waters, in the manufacture of chemicals and drugs, for charging fire extinguishers, and for prevention of timber mold.

Modified Sodas = (So-called Neutral Sodas)

"Modified Sodas" is a term which includes all of those forms of mild alkali which contain more carbonic acid than the normal Sodium Carbonate or Soda Ash. These products are sometimes known as Neutral Sodas.

The Solvay Process Company make the following specialties in Modified Sodas, and are prepared to furnish any other particular combination desired:

Solvay Snow Flake Crystals—Na_2CO_3, $NaHCO_3$, $2H_2O$.

Modified Soda—containing about 27% Bicarbonate of soda, about 60% Carbonate of Soda, making about 45% Actual Alkali (Na_2O).

Modified Soda—containing about 50% Bicarbonate of Soda, about 37% Carbonate of Soda, making about 40% Actual Alkali (Na_2O).

Modified Soda—containing about 64% Bicarbonate of Soda, about 27% Carbonate of Soda, making about 39% Actual Alkali (Na_2O).

Mono-Hydrate Crystals—Na_2CO_3, H_2O.

This is another form of mild alkali, being the normal carbonate with one molecule of water of crystallization.

All of these products are used in cleansing operations, as carried on in laundries, dairies, textile industries, metal cleaning, etc.

See our Special Pamphlets on Snow Flake Crystals and on Metal Cleaning.

For Special Chapter on "Soda as a Cleansing Agent" see Page 24.

PACKAGES: See page 42.

Causticized Ash

Causticized Ash is a term covering mixtures of Soda Ash and Caustic Soda, and is usually graded according to the percentage content of Caustic Soda, i.e., actual NaOH.

We are prepared to make any mixture desired, and have the following grades always in stock:

CAUSTICIZED ASH of 15%, 25%, 36%, 45%, NaOH.

Causticized Ash is used in many cleansing operations where a strong alkali is needed, as in bottle washing and metal cleaning. It is used also for water-softening, and in the manufacture of leather.

See our Special Pamphlets on Metal Cleaning and on Water Purification.

PACKAGES: Barrels, 300 lbs. net.

Crown Filler—$CaSO_4, 2H_2O$

Crown Filler is an extremely pure hydrated sulphate of lime, of a beautiful crystal form. It is the highest grade paper filler known, and is unrivaled by any other filler.

PACKAGES:
> Barrels, 260 lbs., 300 lbs., 370 lbs., net.

Calcium Chloride—$CaCl_2$

Calcium Chloride is furnished as 75% Solid, 75% Granulated, 40% Liquid and 50% Liquid.

It is used as a Refrigeration Brine, for Cold Storage, Air Drying, drying Food Products, laying of Highway Dust, Weed Killing, Prevention of Coal Mine Explosions, in Coal Washing, Tempering of Metals, in the Canning Industry, and for Non-freezing solutions.

See our Special Pamphlets on Calcium Chloride.

PACKAGES:
> 75% Solid, Iron Drums, 610 lbs. net.
> 75% Granulated, Iron Drums, 350, 375 lbs. net.
> Liquid, in tank cars of 4500 gals., 6000 gals. or 10,000 gals. capacity.

Salt—NaCl

The Solvay Process Company's Salt is a fine salt carefully prepared for the trade, and finds extensive use in the textile, leather, and other industries.

PACKAGES:
> Shipped in bulk, carloads and in bags, 200 lbs. net, and barrels, 400 lbs. net.

Limestone—$CaCO_3$

At our extensive quarries of high grade limestone we have installed modern equipment for crushing, sizing and pulverizing limestone.

The crushed limestone is marketed for all concrete and road metal purposes, and is sized for those particular uses.

Solvay Pulverized Limestone for Farm Lands is ground to that degree of fineness required, and represents a superior article for farmers' use.

See our Special Pamphlet on Pulverized Limestone for Farm Lands.

PACKAGES:
> Shipments are made in bulk, in carloads, but the Solvay Pulverized Limestone may be had in 100-lb. bags, if desired, in less than carload lots.

The Valuation of Soda Ash and Caustic Soda as Based on Various Systems of Testing

FROM time to time we have published notes on the various methods of testing and valuing alkalies in this country and abroad, and we again take the opportunity, in this publication, of giving to the trade complete and accurate information on this subject.

Our earlier publications first placed in the hands of consumers the power to ascertain the test by which they buy alkali and we stated that we desired to deal with our customers on the basis of exact analysis. That fact we again emphasize.

We solicit the comparison of our soda with any other make by any of these tests. We guarantee that our soda will compare favorably with any other when both are tested by the same test.

So long as buyers and sellers of soda understand the different tests and know by which one they are buying and selling, no great harm results from these conflicting methods of testing, but when the attempt is made to compare soda by differing tests, confusion and misunderstanding at once arise.

The information contained in the following Tables will be found valuable, both for technical and commercial purposes, and in all dealings which involve the chemical test of Soda Ash and Caustic Soda of all strengths.

The Tables are taken from Lunge's "Hand-book of the Soda Industry." For convenience, the Tables are calculated for even quantities of actual alkali, and extend from the lowest actual alkali tests by differences of $5/10$ of 1% of actual alkali up to the chemically pure product.

Column No. 1 shows the percentage of Sodium Carbonate and of Sodium Hydrate in their respective tables. On the continent of Europe, Soda Ash is generally sold on the content of Sodium Carbonate, and Caustic Soda is sold on the content of Sodium Hydrate (Calculated as Sodium Carbonate).

Column No. 2 shows the percentage of sodium oxide or actual alkali (Na_2O) corresponding to the amount of carbonate or hydrate shown in column No. 1. The actual alkali is reckoned in accordance with the true atomic weights of the elements in the compounds, and is $31/53$ or $62/106$ of the total carbonate of soda.

Column No. 3 shows the amount of sodium oxide or actual alkali (Na_2O) which would be reported by the standard English (Newcastle) test. Under this test, the actual alkali is calculated by the use of an incorrect chemical equivalent for sodium oxide, and is stated as $32/54$ or $64/108$ of the total carbonate of soda. This error originated in the fact that the early chemists fixed the atomic weight

of sodium at 24; subsequent investigations have proved it to be 23. Calculating with the erroneous weight increases the nominal percentage of alkali by 1.3%.

Column No. 4 shows the still more incorrect test which would be reported according to the "New York and Liverpool" method of testing alkali. This test has been in use for the last seventy years, and it is the test by which both soda ash and caustic soda have always been sold in this country. Under this test, the incorrect chemical equivalent for oxide of sodium (Na_2O) is used, while the correct equivalent for carbonate of soda is employed. This test calls the actual alkali 32/53 or 64/106 of the total carbonate of soda, and accordingly gives 3.226% more alkali than actually exists.

In all of our publications in the past, we have pointed out the fact that these different tests of alkali were in more or less general use in different countries, and while it is to be much regretted that no uniform system has been adopted, we have endeavored to fully set forth the differences in order that both buyers and sellers of soda might thoroughly understand the different tests, and know by which one they are buying and selling. If this matter is once fully understood, no possible harm can be done, as the price may be regulated according to the test selected, but a thorough understanding of the matter is desirable, and, as a step in this direction, we present the following tables:

EXAMPLE

N. Y. & Liverpool test gives Na_2O in pure carbonate of soda.... $\dfrac{64}{106} = 60.377\%$

Actual Na_2O is... $\dfrac{62}{106} = 58.491\%$

Difference.................................... 1.886

58.491 : 1.886 :: 100 : 3.226, that is:

The quantity of Na_2O calculated according to the N. Y. & Liverpool test is 3.226% greater than actual Na_2O.

We shall be glad to send on request a leaflet describing in detail standard analytical methods for the testing of alkalies, including methods for the preparation of standard solutions required.

TABLE I ·

For Comparing Different Systems of Alkalimetry for Soda Ash

The following table gives the chemical and commercial equivalents for the different kinds of alkali. On the continent of Europe, alkali is sold by its strength in carbonate of soda (Na_2CO_3), as per column No. 1 of table. In England, alkali is sold nominally on its strength in actual alkali (Na_2O), as per column No. 2 of table, but actually on the so-called "Newcastle Test" of the actual alkali, as per column No. 3 of table. In the United States, the commercial standard for 75 years has been the New York and Liverpool Test for actual alkali, as per column No. 4 of table.

No. 1	No. 2	No. 3	No. 4
Soda Ash Sodium Carbonate Na_2CO_3 Per cent	Actual Alkali Sodium Oxide Na_2O Per cent	Newcastle Test Sodium Oxide Na_2O Per cent	N. Y. & Liv. Sodium Oxide Na_2O Per cent
79.51	46.5	47.11	48.00
80.37	47.0	47.62	48.51
81.22	47.5	48.12	49.03
82.07	48.0	48.63	49.54
82.93	48.5	49.14	50.06
83.78	49.0	49.64	50.58
84.64	49.5	50.15	51.09
85.48	50.0	50.66	51.61
86.34	50.5	51.16	52.12
87.19	51.0	51.67	52.64
88.05	51.5	52.18	53.16
88.90	52.0	52.68	53.67
89.76	52.5	53.19	54.19
90.61	53.0	53.70	54.70
91.47	53.5	54.20	55.22
92.32	54.0	54.71	55.74
93.18	54.5	55.22	56.25
94.03	55.0	55.72	56.77
94.89	55.5	56.23	57.29
95.74	56.0	56.74	57.80
96.60	56.5	57.24	58.32
97.45	57.0	57.75	58.83
98.31	57.5	58.26	59.35
99.16	58.0	58.76	59.87
100.00	58.5	59.27	60.38

TABLE II

For Comparing Different Systems of Alkalimetry for Caustic Soda

Caustic Soda is sold on its strength in Na₂O, as indicated in the New York and Liverpool Test column below.

The price is always based on 60% Caustic, with a proportionate addition for the higher percentages.

No. 1	No. 2	No. 3	No. 4
Caustic Soda Sodium Hydrate NaOH Per cent	Actual Alkali Sodium Oxide Na₂O Per cent	Newcastle Test Sodium Oxide Na₂O Per cent	N. Y. & Liv. Sodium Oxide Na₂O Per cent
74.83	58.0	58.76	59.87
75.48	58.5	59.27	60.38
76.12	59.0	59.77	60.90
76.77	59.5	60.28	61.42
77.40	60.0	60.79	61.93
78.05	60.5	61.30	62.45
78.70	61.0	61.80	62.97
79.35	61.5	62.31	63.48
80.00	62.0	62.82	64.00
80.65	62.5	63.32	64.52
81.29	63.0	63.83	65.03
81.94	63.5	64.33	65.55
82.58	64.0	64.84	66.06
83.23	64.5	65.35	66.58
83.87	65.0	65.85	67.10
84.52	65.5	66.36	67.61
85.16	66.0	66.87	68.13
85.81	66.5	67.37	68.65
86.45	67.0	67.88	69.16
87.10	67.5	68.39	69.68
87.74	68.0	68.89	70.19
88.39	68.5	69.40	70.71
89.03	69.0	69.91	71.23
89.67	69.5	70.41	71.74
90.30	70.0	70.92	72.26
90.95	70.5	71.43	72.77
91.60	71.0	71.93	73.29
92.25	71.5	72.44	73.81
92.90	72.0	72.95	74.32
93.55	72.5	73.45	74.84
94.19	73.0	73.96	75.35
94.84	73.5	74.47	75.87
95.48	74.0	74.97	76.39
96.13	74.5	75.48	76.90
96.77	75.0	75.99	77.42
97.32	75.5	76.49	77.94
98.06	76.0	77.00	78.45
98.71	76.5	77.51	78.97
99.35	77.0	78.01	79.49
100.00	77.5	78.52	80.00

Soda Ash for Glass Making

ONCE upon a time glass was made without Soda Ash, but from the day the Solvay Process lifted Soda Ash out of the class of crude products into the class of pure products, glassmakers have turned to it with always growing enthusiasm as the best, and of late, also the cheapest source of the alkali metal, Sodium.

What the Solvay Process has done for the glassmaker is well shown by what Lomas, the authority of his day, has to say of the Soda Ash of the time just preceding the advent of the Solvay Process: "For the finer sorts of glass and for various other purposes a purer article than Soda Ash is required, and this is readily obtained by dissolving the ash in hot water, settling, boiling down to dryness, and re-furnacing."

Can you imagine treating Solvay Process Soda Ash in that manner; an alkali that is the standard of quality and the purest ingredient of any that enters the glassmaker's batch? It is no wonder that refined alkali under the manufacturing methods used before the perfection of the Solvay Process cost the glassmaker anywhere from three to five times the present price.

There is scarcely another chemical process that requires so nice an adjustment of ingredients as that of glassmaking. This is so for two reasons. First: practically all of the impurities introduced into the melt remain there, and are found in the finished glass. Second: the transparent quality of glass makes any deterioration caused by such impurities perfectly apparent.

It follows, then, that along with the utmost care in selecting sand and lime, must go the greatest vigilance in choosing the Soda Ash. For many years the Solvay Process Co. has made a scientific study of the requirements of the glassmaker, and the result is found in their 58% Dense Soda Ash. The impurities inherent in the product of Lomas' time, referred to above, are practically absent from Solvay Ash. The carbon and sulphides have been entirely eliminated and the iron reduced to the lowest minimum humanly possible. The following average composition of Solvay Process Soda Ash for glassmakers shows the extraordinary purity now attained in this product:

Average analysis of Solvay 58% Dense Soda Ash for entire month of September, 1915—

Sodium Carbonate Na_2CO_3	99.05 %
Sodium Chloride NaCl	.507
Silica SiO_2	.003
Ferric Oxide Fe_2O_3	.005
Aluminum Oxide Al_2O_3	.002
Calcium Carbonate $CaCO_3$	.039
Magnesium Carbonate $MgCO_3$	.030
Sodium Sulphate Na_2SO_4	.055

$$99.691\%$$

Of equal importance with chemical purity are the physical properties of Soda Ash. Experience has shown, for example, that the Ash must be neither too light nor too dense. Formerly, it was thought that the denser the Ash the better. Today it is pretty generally understood that too great a density can be reached. For ease and completeness of melting the mix must be homogeneous, i.e., there must be some Soda Ash right handy for every particle of silica and lime to seize upon. That condition will not be realized if the *bulk* of Soda Ash used is reduced too much by reason of the Ash being too dense. Further, a given method of mixing will not give the same results day after day unless the density of Soda Ash is uniform day after day. Therefore, it is evident that Soda Ash must be not only of the right density, but of uniform density.

Together with the *right* density, and *uniformity* of density must go a non-dusting quality in the Ash, for an Ash that dusts badly is a source of loss, both in handling and in processing, and the dust is a constant irritation to the workmen as well. Dusting is not a function of Density. A dense Ash may dust more badly than a light Ash. The only test is by observation, or better, by making our standard test on dusting which consists in blowing air through the Ash under accurate conditions. It is a simple test and we will be glad to give you the details.

There is no Sodium Glass made that is not the better for having had Soda Ash in the melt. We are not here recommending any particular method or formula for the glassmaker, but it is well known that Soda Ash will lower the fusing point of the melt, give a smoother working glass, and increase the capacity of the plant.

SALT WELLS OF THE SOLVAY PROCESS COMPANY, NEAR SYRACUSE, N. Y.

Caustic Soda for Soap Makers and for Mercerizing

AS in every industry that uses alkali, the business of manufacturing soap has been made easier and more economical by the extraordinary improvement in the last generation in the quality of "Alkali."

From the moment that Ernest Solvay perfected his process for making alkali, known the world over by his name, the way of the soap maker became easier.

The comparison of the caustic soda used a generation ago with the Solvay standard 76% Caustic of today, furnishes a criterion which needs no comment.

	Analysis of White Caustic for Soap in 1881	Analysis of Solvay 76% Caustic, 1916
Sodium Hydrate..................................	83.84%	97.14%
Sodium Carbonate...............................	4.68	1.15
Sodium Chloride.................................	6.52	.85
Sodium Sulphate.................................	4.52	.56
Sodium Sulphide.................................	.025	.00
Sodium Silicate..................................	.463	.372

Whether the soap maker uses caustic soda, or buys soda ash and causticizes it himself, the story is the same. Pure caustic depends on the use of pure soda ash to start with, and, conversely, pure soda ash to start with means a pure finished caustic soda. Our caustic soda is made from soda ash manufactured by the Solvay Process, and the same high degree of technical supervision and the same scientific methods are exercised in producing the caustic soda that are followed in making Solvay soda ash.

Since 1881, the year in which The Solvay Process Company was organized, we have been operating the Solvay process, employing the most skilled technical men that could be found, and giving them the utmost latitude in building up the system of technical factory control, which has resulted in making Solvay alkali products the standard of the world.

During all these thirty-five years of expanding operation, the closest study has been given to the needs of the soap maker. No expense has been spared, first, in finding out just what the trade required in soda ash and caustic, and, second, in giving that product to the trade.

A glance at the analysis of Solvay 76% caustic given above, will show that the deleterious ingredients of the caustic soda of 1881 have been reduced to a minimum, that is, the impurities show a reduction from over 16% to about 3%.

It is essential that caustic soda for soap making be low in salt, sodium sulphate, and sodium carbonate, in order to effect easy solution, and, above all, to insure quick and complete saponification.

In addition to the impurities named, there are often present in commercial caustic soda small quantities of copper. It is known that a very small amount of this impurity will play havoc with the production of a pure soap, for it has been definitely shown that a very minute content of copper will generate heat and decomposition in a neutral soap where there is no excess of alkali. This decomposition results in a darkening in color and in rancidity.

Special methods and the greatest care are required to keep copper out of the finished caustic, and we make a specialty of producing a caustic for soap makers containing the least possible trace of copper.

In all technical manufacture, such as soap making, it is vital that raw material be always of uniform quality. The soap manufacturer is no longer satisfied with turning out a good product three-fourths of the time—he must turn out a perfect product *all* the time. --

For that reason, we lay just as special emphasis on *uniformity* of quality as we do on quality itself, and it is the particular business of one Department to see that each step of the manufacture every hour of the day is progressing as it should. This minute control of every stage of the manufacture results not only in reaching that quality of product desired, but in maintaining it every day of the year.

Mercerizing

For the process of mercerizing only the highest grade caustic soda should be used and particular demand should be made that the content of Sodium Sulphate (Na_2SO_4) and of Sodium Carbonate (Na_2CO_3) be small. These impurities, if present in sufficient amount, will not dissolve in the strong caustic solution required for the process. We recommend for mercerizing our 76% Solvay Caustic Soda, which we carefully prepare for the mercerizing trade.

The impurities named above are also most objectionable in cotton printing, as solid crystals of Sulphate or Carbonate will remain in suspension if low-grade caustic be used and cause much annoyance and trouble.

Soda as a Cleansing Agent

THE use of soda as a direct cleansing agent is of growing importance. In both economy and efficiency it has many advantages over soap in those operations where the foreign matter to be removed consists of oils, fats or greases which are either saponifiable or emulsifiable in solutions of soda, or where the foreign matter is directly soluble in soda solutions and the lubrication afforded by soap is not essential. In many instances the soapy residue, which is always left on articles cleansed with soap, is objectionable, and the use of soda is consequently found much more satisfactory. In addition, much more soap than soda is required to remove a given amount of foreign matter in many cases.

For cleansing purposes, soda ash, caustic soda, and several forms of modified sodas are used. Soda ash and caustic soda are so well known as to need no explanation.

The modified sodas are of three kinds,—first, those containing a mixture of caustic soda and soda ash; second, those containing a mixture of bicarbonate of soda and soda ash; and third, a chemical combination of bicarbonate with soda ash.

Those mixtures containing caustic soda are known as causticized ash, and are placed on the market in mixtures bearing from 10 to 45 per cent. of caustic soda. This class of cleansing soda is intended for use in mechanical cleansing operations in which the operator does not come in contact with the product, and in which caustic soda is too strong an alkali to be used. Among the uses of causticized ash may be mentioned bottle washing for milk, beer, wine, soft-drink and other kinds of bottles in soaking or washing machines. It is used extensively also for cleaning apparatus in the dairy and food products industries where the apparatus is such that it will not be destroyed or attacked by the caustic alkali.

The mixtures or chemical combinations of bicarbonate and soda ash are used in the cleansing of fine textile fabrics, in laundry work, and in cloth finishing. Modified sodas are especially efficient for this work because a maximum cleaning action is obtained with a minimum attack on the goods cleansed. These sodas, properly made, are much more soluble in water than soda ash, and greatly increase the solubility of soap, thereby facilitating the rinsing of soap from goods cleansed with soap. Modified soda mixtures are made in varying proportions to meet the conditions of the operations in which they are to be used.

The modified soda which is a chemical combination of bicarbonate with soda ash is of a special class known as sesqui-carbonate of soda, and is sold by this Company under the name of Solvay Snow Flake Crystals. It has some ad-

vantages over the other forms of modified sodas in that it has a uniform composition, dissolves much more readily, and will not absorb moisture in storage, thus avoiding objectionable caking and hardening. This product has a definite chemical composition of Na_2CO_3, $NaHCO_3$, $2H_2O$.

Mixtures of modified sodas containing a small proportion of bicarbonate are used in the hand-cleansing operations in dairies and bottling works, and in dishwashing machines. This form of soda is especially adapted for cleansing tile and marble floors, which, if cleansed with soap, are left with a darkened appearance and made very slippery. Modified soda of this nature is also very useful for cleansing unfinished wood floors; it does not darken the floor, nor does it collect in the cracks between the boards and become rancid, developing an unpleasant odor, as soap does when used for this purpose.

Both the chemical compound, Solvay Snow Flake Crystals and the bicarbonate mixtures are especially satisfactory in general household cleansing, such as the cleaning of white enameled sinks, sanitary fixtures, refrigerators, washing of dishes, and in general laundry work.

Another product closely related to these modified sodas is mono-hydrate of soda, which is sodium carbonate crystallized with one molecule of water. This is especially adapted to textile cleansing where energetic action is desired, and where the impurities found in commercial soda ash are objectionable.

See our booklet on Metal Cleansing, a practical treatise on Shop Cleaning Methods; also our booklet on "Snow Flake Crystals."

OFFICE OF THE SOLVAY PROCESS CO., DETROIT. MICH. PLANT IN BACKGROUND

Calcium Chloride—Refrigerant

CALCIUM CHLORIDE for ten years past has occupied the dominant place as a refrigerating medium. The cheapness of common salt has been unable to offset its disadvantages, for cheapness in first cost is the only merit possessed by salt as a refrigerant.

Common salt brine corrodes iron pipes so freely that resulting leakage, repairs and delays soon more than offset the low first cost of salt.

In addition, the use of calcium chloride allows much lower temperatures to be used, resulting in higher efficiency all along the line. Smaller volumes are circulated, which results in a saving of power and there is no danger of crystals separating out, thereby reducing the transfer of heat into the cooling medium—upon which depends the efficiency of any system.

Moreover, common salt used for refrigeration or for ice making often contains magnesium chloride, and this accelerates the normal corrosion caused by the sodium chloride. We give on page 30 a table showing the freezing points of common salt brine of different specific gravities, and also the freezing points of Solvay 75% calcium chloride solutions of different specific gravities. From the figures given there, the amount of Solvay calcium chloride required per gallon of water for any desired freezing point may be calculated.

Solvay calcium chloride is *guaranteed to be free from magnesium chloride* and to contain 73% to 75% calcium chloride.

Freedom from magnesium chloride is of prime importance. With magnesium chloride present, a leakage of ammonia into the brine means a precipitation of magnesium hydrate with resulting scaling and clogging of pipes. It means also the formation of ammonium chloride, or sal-ammoniac, which is strongly corrosive, especially in the presence of air. On the other hand, a leakage of ammonia into pure calcium chloride produces no effect, as ammonia is soluble in a pure calcium chloride solution.

The following analyses represent the average composition of Solvay 75% calcium chloride and calcium magnesium chlorides found on the market:

	Solvay 75% Calcium Chloride	Calcium Magnesium Chlorides	
		A	B
Calcium Chloride, $CaCl_2$	73.59	43.70	46.26
Sodium Chloride (Salt), NaCl	1.45	1.88	1.46
Magnesium Chloride, $MgCl_2$	0.00	18.80	18.98
Insoluble in water	.07	.48	.20
Total Solids	75.11	64.86	66.90
Water	24.89	35.14	33.10
Total	100.00	100.00	100.00

On page 30 is given a table, showing the freezing points obtained with equal weights of Solvay 75% calcium chloride and calcium magnesium chloride, of which the analysis is given above, together with the increased quantity of the calcium magnesium chloride required to produce the same freezing point as that of Solvay calcium chloride brine.

Solvay 75% calcium chloride gives the same freezing point with 10% to 15% less weight of calcium chloride per gallon. In other words, 100 lbs. will do the work of 110 to 115 lbs. of calcium magnesium chloride.

By the continued action of air, any refrigerating brine will finally become acid and this, of course, greatly increases its corrosive properties, and a feature of Solvay calcium chloride is the ease with which an acid condition can be corrected. This is done by simply hanging a bag containing a few lumps of slaked lime in the brine tank, preferably near the return pipe.

A brine *containing magnesium chloride* cannot be corrected in this way, as lime would precipitate magnesium hydrate.

We have here noted only the salient points which have given true calcium chloride its pre-eminent place as a refrigerant. For a more extended treatment of the subject, we refer those interested to our booklet "Solvay 75% Calcium Chloride," which will be gladly sent on request.

Solvay Granulated Calcium Chloride

"A Natural Dust Layer"

This product is a clean chemical salt, odorless and colorless and without effect on rubber.

These properties, together with its capacity to absorb moisture from the air, make it a highly effective and desirable dust-layer for road work. If interested, please write for our booklet, "Solvay Granulated Calcium Chloride."

PLANT OF THE SOLVAY PROCESS CO., SYRACUSE, N. Y., 1886

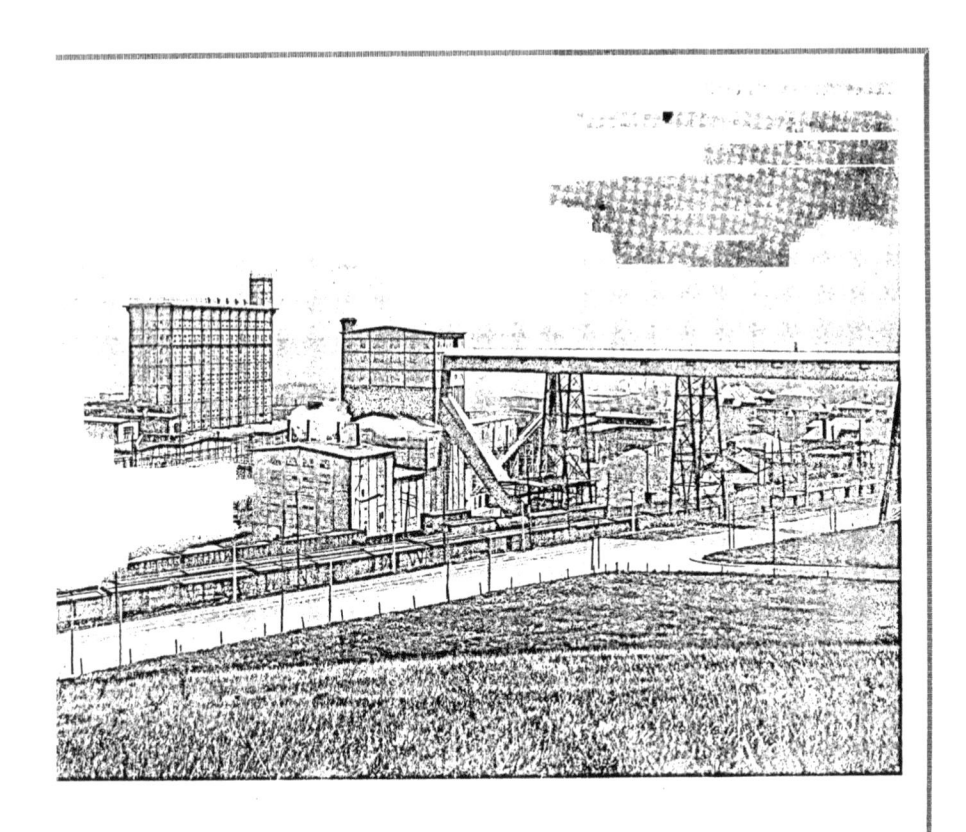

PLANT OF
THE SOLVAY PROCESS COMPANY
SYRACUSE, N. Y.
1916

TABLE III
Common Salt Brine at 60° Fahrenheit

Degrees Baumé	Degrees Salometer	Specific Gravity	Per Cent. Salt	Weight of One Gallon	Weight of One Cubic Foot	Freezing Point
5	20	1037	5	8.7	64.7	25.4° F.
10	40	1073	10	9.0	67.0	18.6
15	60	1115	15	9.3	69.6	12.2
19	80	1150	20	9.6	71.8	6.9
23	100	1191	25	9.9	74.3	1.0

The following table gives the strength and freezing points of Solvay 75% Calcium Chloride Solutions:—

Sp. Gr. @ 68° F.	Lbs. 75% Solvay Cal. Chl. per Gal.	Lbs. 75% Solvay Cal. Chl. per Cu. Ft.	Freezing Point ° Fahr.
1100	1.46	10.9	+18.0
1125	1.83	13.7	+12.5
1150	2.20	16.5	+ 6.5
1175	2.59	19.4	— 2.0
1200	2.99	22.4	—12.5
1225	3.38	25.3	—23.5
1250	3.75	28.3	—36.5

TABLE IV
Freezing Points of Brine Made with Equal Weights of Solvay and of Calcium Magnesium Chlorides

	Lbs. per Gal.	Sp. Gr. at 65° F.	Freezing Point
Solvay 75% Calcium Chloride................	3.0	1202	—12.5° F.
Calcium Magnesium Chloride, A..............	3.0	1175	—6.0°
Calcium Magnesium Chloride, B.............	3.0	1174	—7.0°
Calcium Magnesium Chloride, A, to give same freezing point as Solvay..................	3.27	1189	—12.5°
Calcium Magnesium Chloride, B, to give same freezing point as Solvay..................	3.27	1190	—12.5°
Solvay 75% Calcium Chloride................	3.5	1233	—28°
Calcium Magnesium Chloride, A..............	3.5	1199	—19°
Calcium Magnesium Chloride, B..............	3.5	1199	—19°
Calcium Magnesium Chloride, A, to give same freezing point as Solvay..................	3.96	1220	—28°
Calcium Magnesium Chloride, B, to give same freezing point as Solvay..................	3.89	1222	—28°

TABLE V
Solubility of Soda Salts

Temperature		SODIUM CARBONATE DRY Na₂CO₃		SODIUM MONOHYDRATE Na₂CO₃.H₂O		SAL SODA Na₂CO₃.10H₂O		SODIUM BI-CARBONATE NaHCO₃		CAUSTIC SODA NaOH	
°C	°F	Parts Per 100 Parts Water	Parts In 100 Parts Solution	Parts Per 100 Parts Water	Parts In 100 Parts Solution	Parts Per 100 Parts Water	Parts In 100 Parts Solution	Parts Per 100 Parts Water	Parts In 100 Parts Solution	Parts Per 100 Parts Water	Parts In 100 Parts Sol.
0°	32°	7.1	6.6	8.3	7.6	19.2	16.1	6.9	6.5		
5	41	9.5	8.7	11.1	10.0	25.7	20.5	7.5	7.0		
10	50	12.6	11.2	14.7	12.8	34.0	25.4	8.2	7.5		
15	59	16.5	14.2	19.3	16.2	44.5	30.9	8.9	8.2		
20	68	21.5	17.7	25.2	20.1	58.1	36.7	9.6	8.8	109	52.2
25	77	28.2	22.0	33.0	24.8	76.1	43.2	10.4	9.4		
30	86	37.8	27.4	44.2	30.7	102.1	50.5	11.1	10.0	119	54.3
32.5	90.5	46.2	31.6	54.1	35.1	124.7	55.5				
35	95	46.2	31.6	54.1	35.1	124.7	55.5	11.9	10.6		
40	104	46.1	31.5	53.9	35.0	124.5	55.4	12.7	11.3	129	56.3
60	140	46.0	31.5	53.8	35.0	124.2	55.4	16.4	13.8	174	63.5
80	176	45.8	31.4	53.6	34.8	123.7	55.3			313	75.8
100	212	45.5	31.3	53.2	34.7	122.9	55.1				
105	221	45.2	31.1	52.9	34.6	122.0	55.0				
110	230									365	78.5
192	378									521	83.9

Note: Figures for Sodium Carbonate taken from "Solubilities of Inorganic and Organic Substances" by Seidell, p. 296. Figures for Sodium Hydrate from Pickering, Jour. Chem. Soc., 63, 890, 1893.

The solubility of sodium carbonate in water increases from 0° C. up to a temperature somewhere between 31° C. and 35° C., at which point it becomes practically constant.

When sodium carbonate is dissolved in water, various hydrates may be formed. Ketner, in Zeitschrift für Physikalische Chemie, Vol. 39, page 645, states that up to a temperature of 31.85° C. the carbonate exists in solution as Na₂CO₃,10H₂O; between 31.85° and 35.1° it exists as Na₂CO₃,7H₂O, and above 35.1 as Na₂CO₃,H₂O.

The above table shows the solubility of dry Na₂CO₃ in water, and sodium carbonate calculated as mono-hydrate and deca-hydrate. The two hydrates are shown for all temperatures of the table, although the mono-hydrate does not actually exist in solution at the lower temperatures, and the deca-hydrate does not exist at the higher temperatures.

The solubility of Sodium Hydrate below 15° or 20° C. is dependent on the particular hydrate formed, and since figures for those low temperatures are not commercially important, they are here omitted.

TABLE VI

Specific Gravity of Solutions of Sodium Carbonate

AT 15° C. (59° F.)

Specific Gravity	Degrees Baumé	Degrees Twaddle	Per cent by Weight		1 Litre Contains Grams		100 Gallons Contain Pounds	
			Na₂CO₃	Na₂CO₃+ 10H₂O	Na₂CO₃	Na₂CO₃+ 10H₂O	Na₂CO₃	Na₂CO₃+ 10H₂O
1.007	1	1.4	0.67	1.81	6.8	18.2	5.6	15.2
1.014	2	2.8	1.33	3.59	13.5	36.4	11.3	30.4
1.022	3	4.4	2.09	5.64	21.4	57.6	17.8	48.0
1.029	4	5.8	2.76	7.44	28.4	76.6	23.7	63.9
1.036	5	7.2	3.43	9.25	35.5	95.8	29.6	79.9
1.045	6	9.0	4.29	11.6	44.8	120.9	37.3	100.8
1.052	7	10.4	4.94	13.3	52.0	140.2	43.4	116.9
1.060	8	12.0	5.71	15.4	60.5	163.2	50.5	136.1
1.067	9	13.4	6.37	17.2	68.0	183.3	56.7	152.9
1.075	10	15.0	7.12	19.2	76.5	206.4	63.8	172.1
1.083	11	16.6	7.88	21.3	85.3	230.2	71.1	192.0
1.091	12	18.2	8.62	23.2	94.0	253.6	78.4	211.5
1.100	13	20.0	9.43	25.4	103.7	279.8	86.5	233.4
1.108	14	21.6	10.2	27.5	112.9	304.5	94.2	253.9
1.116	15	23.2	10.9	29.5	122.2	329.6	101.9	274.8
1.125	16	25.0	11.8	31.9	132.9	358.3	110.8	298.8
1.134	17	26.8	12.6	34.0	143.0	385.7	119.3	321.7
1.142	18	28.4	13.5	35.5	150.3	405.3	125.4	338.0
1.152	19	30.4	14.2	38.4	164.1	442.4	136.9	369.0

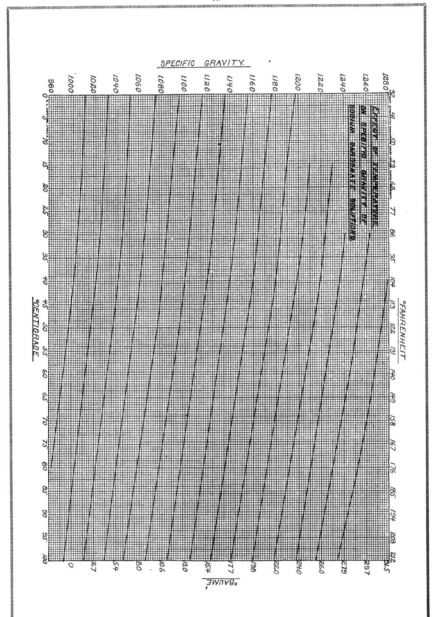

SPECIFIC GRAVITY.

EFFECT OF TEMPERATURE
ON SPECIFIC GRAVITY OF
SODIUM CARBONATE SOLUTIONS.

°FAHRENHEIT

°CENTIGRADE.

°BAUMÉ.

TABLE VII

Specific Gravity of Solutions of Pure Caustic Soda

AT 15° C. (59° F.)

Specific Gravity	Degrees Baumé	Degrees Twaddle	Per cent Actual Alkali Na_2O	Per cent Caustic Soda NaOH	1 Litre Contains Grams		100 Gallons Contain Pounds	
					Actual Alkali Na_2O	Caustic Soda NaOH	Actual Alkali Na_2O	Caustic Soda NaOH
1.007	1	1.4	0.47	0.61	4	6	3.34	5.00
1.014	2	2.8	0.93	1.20	9	12	7.51	10.00
1.022	3	4.4	1.55	2.00	16	21	13.34	17.51
1.029	4	5.8	2.10	2.71	22	28	18.35	23.35
1.036	5	7.2	2.60	3.35	27	35	22.52	29.19
1.045	6	9.0	3.10	4.00	32	42	26.69	35.03
1.052	7	10.4	3.60	4.64	38	49	31.69	40.87
1.060	8	12.0	4.10	5.29	43	56	35.86	46.70
1.067	9	13.4	4.55	5.87	49	63	40.87	52.54
1.075	10	15.0	5.08	6.55	55	70	45.87	58.38
1.083	11	16.6	5.67	7.31	61	79	50.87	65.89
1.091	12	18.2	6.20	8.00	68	87	56.71	72.56
1.100	13	20.0	6.73	8.68	74	95	61.72	79.23
1.108	14	21.6	7.30	9.42	81	104	67.55	86.74
1.116	15	23.2	7.80	10.06	87	112	72.56	93.40
1.125	16	25.0	8.50	10.97	96	123	80.06	102.58
1.134	17	26.8	9.18	11.84	104	134	86.74	111.76
1.142	18	28.4	9.80	12.64	112	144	93.41	120.10
1.152	19	30.4	10.50	13.55	121	156	100.91	130.10
1.162	20	32.4	11.14	14.37	129	167	107.59	139.28
1.171	21	34.2	11.73	15.13	137	177	114.26	147.62
1.180	22	36.0	12.33	15.91	146	188	121.76	156.79
1.190	23	38.0	13.00	16.77	155	200	129.27	166.80
1.200	24	40.0	13.70	17.67	164	212	136.78	176.81
1.210	25	42.0	14.40	18.58	174	225	145.12	187.65

TABLE VII (Continued)

Specific Gravity of Solutions of Pure Caustic Soda

AT 15° C. (59° F.)

Specific Gravity	Degrees Baumé	Degrees Twaddle	Per cent Actual Alkali Na$_2$O	Per cent Caustic Soda NaOH	1 Litre Contains Grams		100 Gallons Contain Pounds	
					Actual Alkali Na$_2$O	Caustic Soda NaOH	Actual Alkali Na$_2$O	Caustic Soda NaOH
1.220	26	44.0	15.18	19.58	185	239	154.29	199.33
1.231	27	46.2	15.96	20.59	196	253	163.46	211.00
1.241	28	48.2	16.76	21.42	208	266	173.47	221.84
1.252	29	50.4	17.55	22.64	220	283	183.48	236.02
1.263	30	52.6	18.35	23.67	232	299	193.49	249.37
1.274	31	54.8	19.23	24.81	245	316	204.33	263.54
1.285	32	57.0	20.00	25.80	257	332	214.34	276.88
1.297	33	59.4	20.80	26.83	270	348	225.18	290.23
1.308	34	61.6	21.55	27.80	282	364	235.19	303.58
1.320	35	64.0	22.35	28.83	295	381	246.03	317.75
1.332	36	66.4	23.20	29.93	309	399	257.71	332.77
1.345	37	69.0	24.20	31.22	326	420	271.88	350.28
1.357	38	71.4	25.17	32.47	342	441	285.23	367.79
1.370	39	74.0	26.12	33.69	359	462	299.41	385.31
1.383	40	76.6	27.10	34.96	375	483	312.75	402.82
1.397	41	79.4	28.10	36.25	392	506	326.92	422.00
1.410	42	82.0	29.05	37.47	410	528	341.94	440.35
1.424	43	84.8	30.08	38.80	428	553	356.95	461.20
1.438	44	87.6	31.00	39.99	446	575	371.96	479.55
1.453	45	90.6	32.10	41.41	466	602	388.64	502.07
1.468	46	93.6	33.20	42.83	487	629	406.16	524.58
1.483	47	96.6	34.40	44.38	510	658	425.34	548.77
1.498	48	99.6	35.70	46.15	535	691	446.19	576.29
1.514	49	102.8	36.90	47.60	559	721	466.21	601.31
1.530	50	106.0	38.00	49.02	581	750	484.55	625.50

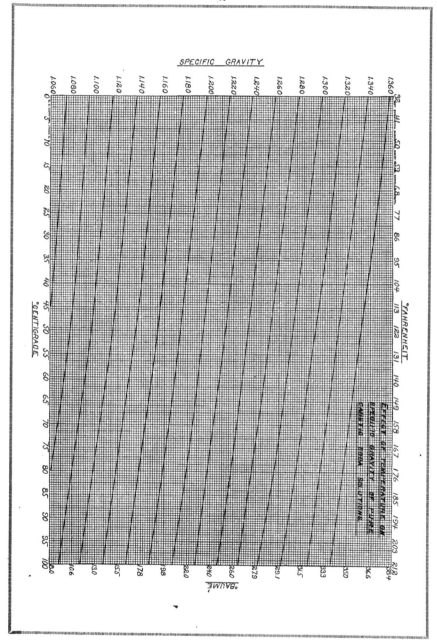

VIEW IN THE CHEMICAL RESEARCH LABORATORY, SYRACUSE, N. Y.

A CORNER OF THE CHEMICAL RESEARCH LABORATORY, SYRACUSE, N. Y.

TABLE VIII

Equivalent Prices for Soda Ash (Basis 48%)

Price per 100 lbs. 48 Per Cent.	Price per 100 lbs. 58 Per Cent.	Price per Ton. (2,240 lbs.) 48 Per Cent.	Price per Ton. (2,240 lbs.) 58 Per Cent.	Price per 1000 Kilos. (2,204.6 lbs.) 48 Per Cent.	Price per 1000 Kilos. (2,204.6 lbs.) 58 Per Cent.
$ .01	$ .0121	$ .224	$ .2706	$ .2205	$ .2664
.02	.0242	.448	.5413	.4409	.5328
.03	.0362	.672	.8120	.6614	.7992
.04	.0483	.896	1.0826	.8818	1.0656
.05	.0604	1.120	1.3533	1.1023	1.3319
.06	.0725	1.344	1.6240	1.3228	1.5983
.07	.0846	1.568	1.8946	1.5432	1.8647
.08	.0966	1.792	2.1653	1.7637	2.1311
.09	.1087	2.016	2.4360	1.9841	2.3975
.10	.1208	2.240	2.7066	2.2046	2.6639
.15	.1812	3.360	4.0600	3.3069	3.9958
.20	.2417	4.480	5.4133	4.4092	5.3278
.25	.3021	5.600	6.7666	5.5115	6.6597
.30	.3625	6.720	8.1200	6.6138	7.9917
.35	.4229	7.840	9.4733	7.7161	9.3236
.40	.4833	8.960	10.8266	8.8184	10.6556
.45	.5437	10.080	12.1800	9.9207	11.9875
.50	.6042	11.200	13.5333	11.0230	13.3194
.55	.6646	12.320	14.8866	12.1253	14.6514
.60	.7250	13.440	16.2400	13.2276	15.9833
.65	.7854	14.560	17.5933	14.3299	17.3153
.70	.8458	15.680	18.9466	15.4322	18.6472
.75	.9062	16.800	20.3000	16.5345	19.9792
.80	.9666	17.920	21.6533	17.6368	21.3111
.85	1.0271	19.040	23.0066	18.7391	22.6431
.90	1.0875	20.160	24.3600	19.8415	23.9750
.95	1.1479	21.280	25.7133	20.9437	25.3070
1.00	1.2083	22.400	27.0666	22.0460	26.6389
1.05	1.2687	23.520	28.4200	23.1483	27.9708
1.10	1.3292	24.640	29.7733	24.2506	29.3028
1.15	1.3896	25.760	31.1266	25.3529	30.6348
1.20	1.4500	26.880	32.4800	26.4552	31.9667
1.25	1.5104	28.000	33.8333	27.5575	33.2986
1.30	1.5708	29.120	35.1866	28.6598	34.6306
1.35	1.6312	30.240	36.5400	29.7621	35.9625

TABLE IX
Equivalent Prices for Caustic Soda, per 100 Lbs.
(Basis 60%)

Price per 100 lbs. 60 Per Cent.	Price per 100 lbs. 70 Per Cent.	Price per 100 lbs. 74 Per Cent.	Price per 100 lbs. 76 Per Cent.
$ .01	$ .0117	$ .0123	$ .0127
.02	.0233	.0246	.0253
.03	.0350	.0370	.0380
.04	.0467	.0493	.0507
.05	.0583	.0617	.0633
.06	.0700	.0740	.0760
.07	.0817	.0863	.0887
.08	.0933	.0987	.1013
.09	.1050	.1100	.1140
.10	.1167	.1233	.1267
.15	.1750	.1850	.1900
.20	.2333	.2467	.2533
.25	.2916	.3083	.3167
.30	.3500	.3700	.3800
.40	.4667	.4933	.5067
.50	.5833	.6167	.6333
.60	.7000	.7400	.7600
.70	.8167	.8633	.8867
.80	.9333	.9867	1.0133
.90	1.0500	1.1100	1.1400
1.00	1.1667	1.2333	1.2667
1.10	1.2833	1.3567	1.3933
1.20	1.4000	1.4800	1.5200
1.30	1.5167	1.6033	1.6467
1.40	1.6333	1.7267	1.7733
1.50	1.7500	1.8500	1.9000
1.60	1.8667	1.9733	2.0267
1.70	1.9833	2.0967	2.1533
1.80	2.1000	2.2200	2.2800
1.90	2.2167	2.3433	2.4067
2.00	2.3333	2.4667	2.5333
2.10	2.4500	2.5900	2.6600
2.20	2.5667	2.7133	2.7867
2.30	2.6833	2.8367	2.9133
2.40	2.8000	2.9600	3.0400
2.50	2.9167	3.0833	3.1667
2.60	3.0333	3.2067	3.2933
2.70	3.1500	3.3300	3.4200
2.80	3.2667	3.4533	3.5467
2.90	3.3833	3.5767	3.6733
3.00	3.5000	3.7000	3.8000

TABLE X

Equivalent Prices for Caustic Soda by Tons (Basis 60%)

Price per 100 lbs. 60 Per Cent.	Price per Ton. (2,240 lbs.) 60 Per Cent.	Price per Ton. (2,240 lbs.) 70 Per Cent.	Price per Ton. (2,240 lbs.) 74 Per Cent.	Price per Ton. (2,240 lbs.) 76 Per Cent.
$ .01	$ .224	$ .261	$ .276	$ .284
.02	.448	.523	.552	.567
.03	.672	.784	.829	.851
.04	.896	1.045	1.105	1.135
.05	1.120	1.307	1.381	1.419
.06	1.344	1.568	1.658	1.702
.07	1.568	1.829	1.934	1.986
.08	1.792	2.091	2.220	2.270
.09	2.016	2.352	2.486	2.554
.10	2.240	2.613	2.764	2.837
.15	3.360	3.920	4.144	4.256
.20	4.480	5.227	5.525	5.675
.25	5.600	6.533	6.907	7.093
.30	6.720	7.840	8.288	8.512
.40	8.960	10.453	11.051	11.349
.50	11.200	13.068	13.813	14.187
.60	13.440	15.680	16.576	17.024
.70	15.680	18.293	19.338	19.861
.80	17.920	20.907	22.100	22.698
.90	20.160	23.520	24.864	25.536
1.00	22.400	26.133	27.626	28.373
1.10	24.640	28.747	30.389	31.210
1.20	26.880	31.360	33.152	34.048
1.30	29.120	33.973	35.914	36.885
1.40	31.360	36.587	38.677	39.722
1.50	33.600	39.200	41.440	42.560
1.60	35.840	41.813	44.202	45.397
1.70	38.080	44.427	46.965	48.234
1.80	40.320	47.040	49.728	51.072
1.90	42.560	49.653	52.490	53.909
2.00	44.800	52.267	55.253	56.747
2.10	47.040	54.880	58.016	59.584
2.20	49.280	57.493	60.777	62.421
2.30	51.520	60.106	63.540	65.258
2.40	53.760	62.720	66.304	68.096
2.50	56.000	65.333	69.067	70.933
2.60	58.240	67.946	71.828	73.770
2.70	60.480	70.560	74.594	76.607
2.80	62.720	73.173	77.355	79.445
2.90	64.960	75.787	80.117	82.282
3.00	67.200	78.400	82.880	85.120

TABLE XI
Equivalent Prices for Caustic Soda by Kilos (Basis 60%)

Price per 100 lbs. 60 Per Cent.	Price per 1000 Kilos. (2,204.6 lbs.) 60 Per Cent.	Price per 1000 Kilos. (2,204.6 lbs.) 70 Per Cent.	Price per 1000 Kilos. (2,204.6 lbs.) 74 Per Cent.	Price per 1000 Kilos. (2,204.6 lbs.) 76 Per Cent.
$ .01	$ .2205	$ .2572	$ .2719	$ .2793
.02	.4409	.5144	.5438	.5585
.03	.6614	.7716	.8157	.8378
.04	.8818	1.0288	1.0876	1.1170
.05	1.1023	1.2860	1.3595	1.3963
.06	1.3228	1.5432	1.6314	1.6756
.07	1.5432	1.8004	1.9033	1.9547
.08	1.7637	2.0576	2.1752	2.2340
.09	1.9841	2.3148	2.4471	2.5132
.10	2.2046	2.5720	2.7190	2.7925
.15	3.3069	3.8580	4.0785	4.1888
.20	4.4092	5.1440	5.4380	5.5849
.25	5.5115	6.4301	6.7975	6.9812
.30	6.6138	7.7161	8.1570	8.3775
.40	8.8184	10.2881	10.8760	11.1700
.50	11.0230	12.8602	13.5950	13.9625
.60	13.2276	15.4322	16.3140	16.7549
.70	15.4322	18.0042	19.0330	19.5474
.80	17.6368	20.5763	21.7520	22.3400
.90	19.8414	23.1483	24.4710	25.1325
1.00	22.0460	25.7203	27.1901	27.9250
1.10	24.2506	28.2924	29.9091	30.7174
1.20	26.4552	30.8644	32.6281	33.5099
1.30	28.6598	33.4364	35.3471	36.2025
1.40	30.8644	36.0085	38.0661	39.0949
1.50	33.0690	38.5805	40.7851	41.8880
1.60	35.2736	41.1525	43.5041	44.6799
1.70	37.4782	43.7246	46.2231	47.4723
1.80	39.6828	46.2966	48.9421	50.2649
1.90	41.8874	48.8686	51.6611	53.0574
2.00	44.0920	51.4407	54.3801	55.8490
2.10	46.2966	54.0127	57.0991	58.6423
2.20	48.5012	56.5847	59.8181	61.4348
2.30	50.7058	59.1568	62.5372	64.2274
2.40	52.9104	61.7288	65.2562	67.0199
2.50	55.1150	64.3008	67.9752	69.8123
2.60	57.3196	66.8729	70.6942	72.6048
2.70	59.5242	69.4449	73.4132	75.3973
2.80	61.7288	72.0169	76.1322	78.1899
2.90	63.9334	74.5890	78.8512	80.9823
3.00	66.1380	77.1610	81.5702	83.7748

TABLE XII

Summary of Shipping Weights of Solvay Products

Net Weight — Pounds

Product	Bags	Barrels	Drums	Bulk Lbs. per Cu. Ft.
58% Soda Ash—Light............	300	300 (Approx.)		31¼—34⅓
58% Soda Ash—Dense...........	500	500 (Approx.)		64—67
48% Soda Ash—Ordinary.......	300	300 (Approx.)		34½—37
48% Soda Ash—Special.........	500			64—66
Caustic Soda—Solid............			675	
Caustic Soda—Ground..........		550, 575	400	
Solvay Snow Flake Crystals......		280, 350		
Modified Sodas.................		280		
Mono-Hydrate Crystals.........		450		
Causticized Ash................		300		
Crown Filler...................		260, 300, 370		
Calcium Chloride—Solid........			610	
Calcium Chloride—Granulated....			350, 375	
Salt..........................	200	400		60
Limestone—Pulverized..........	100			75

DETROIT PLANT OF THE SOLVAY PROCESS CO.

TABLE XIII

Chemical Equivalents of Solvay Products

Name	Molecular Formula	Molecular Weight	Percentage Composition		
Sodium Carbonate (Soda Ash)	Na_2CO_3	106	Na_2O CO_2	58.49 41.51	 100
Sodium Carbonate (Crystal)	Na_2CO_3+ $10H_2O$	286	Na_2O CO_2 H_2O	21.68 15.39 62.93	 100
Sodium Bicarbonate	$NaHCO_3$	84	Na_2O CO_2 H_2O	36.90 52.38 10.72	 100
Sodium Monohydrate Crystals	Na_2CO_3+ H_2O	124	Na_2O CO_2 H_2O	50.00 35.48 14.52	 100
Sodium Sesquicarbonate Snow Flake Crystals	Na_2CO_3 $NaHCO_3+$ $2H_2O$	226	Na_2O CO_2 H_2O H_2O	41.15 38.94 15.93 (Crys.) 3.98 (Comb.)	 100
Sodium Sulphate	Na_2SO_4	142	Na_2O SO_3	43.66 56.34	 100
Sodium Sulphate Crystal	Na_2SO_4+ $10H_2O$	322	Na_2O SO_3 H_2O	19.25 24.84 55.91	 100
Sodium Hydrate (Caustic)	$NaOH$	40	Na_2O H_2O	77.50 22.50	 100
Sodium Oxide	Na_2O	62	Na O	74.19 25.81	 100
Sodium Chloride	$NaCl$	58.5	Na Cl	39.32 60.68	 100
Calcium Chloride	$CaCl_2$	111	Ca Cl	36.04 63.96	 100
Calcium Carbonate	$CaCO_3$	100	CaO CO_2	56.0 44.0	 100
Calcium Oxide (Caustic Lime)	CaO	56	Ca O	71.43 28.57	 100
Calcium Sulphate (dry)	$CaSO_4$	136	CaO SO_3	41.18 58.82	 100
Calcium Sulphate (Crystal) Crown Filler	$CaSO_4+$ $2H_2O$	172	CaO SO_3 H_2O	32.56 46.51 20.93	 100
Calcium Hydroxide (Hydrate of Lime)	$Ca(OH)_2$	74	CaO H_2O	75.67 24.33	 100

TABLE XIV

Comparison of Fahrenheit and Centigrade Temperatures with Centigrade Degrees as Basis

To use the Table, find Centigrade temperature in intervals of "tens" in left hand column, move across table to point under proper "unit." Figure found at intersection is the corresponding Fahrenheit Temperature.

Note: For Comparison of Temperatures between -1° C. and -9° C., see small table, above main table.

—1° C.	—2° C.	—3° C.	—4° C.	—5° C.	—6° C.	—7° C.	—8° C.	—9° C.
+30.2° F.	+28.4° F.	+26.6° F.	+24.8° F.	+23.0° F.	+21.2° F.	+19.4° F.	+17.6° F.	+15.8° F.

° C.		1	2	3	4	5	6	7	8	9
	° F.	° F.	° F.	° F.	° F.	° F.	° F.	° F.	° F.	° F.
— 40	—40	—41.8	—43.6	—45.4	—47.2	—49	—50.8	—52.6	—54.4	—56.2
— 30	—22	—23.8	—25.6	—27.4	—29.2	—31	—32.8	—34.6	—36.4	—38.2
— 20	— 4	— 5.8	— 7.6	— 9.4	—11.2	—13	—14.8	—16.6	—18.4	—20.2
— 10	+14	+12.2	+10.4	+ 8.6	+ 6.8	+ 5	+ 3.2	+ 1.4	+ 0.4	— 2.2
0	32	+33.8	35.6	37.4	39.2	41	42.8	44.6	46.4	48.2
+ 10	50	51.8	53.6	55.4	57.2	59	60.8	62.6	64.4	66.2
+ 20	68	69.8	71.6	73.4	75.2	77	78.8	80.6	82.4	84.2
+ 30	86	87.8	89.6	91.4	93.2	95	96.8	98.6	100.4	102.2
+ 40	104	105.8	107.6	109.4	111.2	113	114.8	116.6	118.4	120.2
+ 50	122	123.8	125.6	127.4	129.2	131	132.8	134.6	136.4	138.2
+ 60	140	141.8	143.6	145.4	147.2	149	150.8	152.6	154.4	156.2
+ 70	158	159.8	161.6	163.4	165.2	167	168.8	170.6	172.4	174.2
+ 80	176	177.8	179.6	181.4	183.2	185	186.8	188.6	190.4	192.2
+ 90	194	195.8	197.6	199.4	201.2	203	204.8	206.6	208.4	210.2
+100	212	213.8	215.6	217.4	219.2	221	222.8	224.6	226.4	228.2
+110	230	231.8	233.6	235.4	237.2	239	240.8	242.6	244.4	246.2
+120	248	249.8	251.6	253.4	255.2	257	258.8	260.6	262.4	264.2
+130	266	267.8	269.6	271.4	273.2	275	276.8	278.6	280.4	282.2
+140	284	285.8	287.6	289.4	291.2	293	294.8	296.6	298.4	300.2
+150	302	303.8	305.6	307.4	309.2	311	312.8	314.6	316.4	318.2
+160	320	321.8	323.6	325.4	327.2	329	330.8	332.6	334.4	336.2

General formulas for converting Fahrenheit temperatures to corresponding Centigrade temperatures, and vice versa:

If c and f denote corresponding temperatures on the Centigrade and Fahrenheit scales, respectively, then:

$$c = 5/9 \ (f-32) \text{ and } f = 9/5 \ c + 32.$$

From these equations it follows: To convert Fahrenheit temperatures to Centigrade temperatures, subtract 32 and multiply by 5/9.

Examples: $104° \text{ F} = (104-32) \times 5/9 = 40° \text{ C.}$
$-31° \text{ F} = (-31-32) \times 5/9 = -35° \text{ C.}$

To convert Centigrade temperatures to Fahrenheit temperatures, multiply by 9/5 and add 32:

Examples: $10° \text{ C.} = (10 \times 9/5) + 32 = 50° \text{ F.}$
$-36° \text{ C.} = (-36 \times 9/5) + 32 = -32.8° \text{ F.}$

TABLE XV

International Atomic Weights—1916

	Symbol	Atomic Weight		Symbol	Atomic Weight
Aluminium.........	Al	27.1	Neodymium........	Nd	144.3
Antimony...........	Sb	120.2	Neon..............	Ne	20.2
Argon..............	A	39.88	Nickel.............	Ni	58.68
Arsenic.............	As	74.96	Niton (radium emana-		
Barium.............	Ba	137.37	tion).............	Nt	222.4
Bismuth...........	Bi	208.0	Nitrogen...........	N	14.01
Boron..............	B	11.0	Osmium............	Os	190.9
Bromine...........	Br	79.92	Oxygen............	O	16.00
Cadmium...........	Cd	112.40	Palladium..........	Pd	106.7
Caesium............	Cs	132.81	Phosphorus........	P	31.04
Calcium............	Ca	40.07	Platinum..........	Pt	195.2
Carbon.............	C	12.005	Potassium..........	K	39.10
Cerium.............	Ce	140.25	Praseodymium......	Pr	140.9
Chlorine...........	Cl	35.46	Radium............	Ra	226.0
Chromium..........	Cr	52.0	Rhodium..........	Rh	102.9
Cobalt.............	Co	58.97	Rubidium..........	Rb	85.45
Columbium.........	Cb	93.5	Ruthenium........	Ru	101.7
Copper.............	Cu	63.57	Samarium.........	Sa	150.4
Dysprosium........	Dy	162.5	Scandium..........	Sc	44.1
Erbium.............	Er	167.7	Selenium..........	Se	79.2
Europium..........	Eu	152.0	Silicon.............	Si	28.3
Fluorine............	F	19.0	Silver.............	Ag	107.88
Gadolinium........	Gd	157.3	Sodium............	Na	23.00
Gallium............	Ga	69.9	Strontium.........	Sr	87.63
Germanium.........	Ge	72.5	Sulphur...........	S	32.06
Glucinum..........	Gl	9.1	Tantalum..........	Ta	181.5
Gold...............	Au	197.2	Tellurium..........	Te	127.5
Helium.............	He	4.00	Terbium...........	Tb	159.2
Holmium...........	Ho	163.5	Thallium..........	Tl	204.0
Hydrogen..........	H	1.008	Thorium...........	Th	232.4
Indium.............	In	114.8	Thulium...........	Tm	168.5
Iodine.............	I	126.92	Tin...............	Sn	118.7
Iridium............	Ir	193.1	Titanium..........	Ti	48.1
Iron...............	Fe	55.84	Tungsten..........	W	184.0
Krypton...........	Kr	82.92	Uranium..........	U	238.2
Lanthanum.........	La	139.0	Vanadium.........	V	51.0
Lead...............	Pb	207.20	Xenon.............	Xe	130.2
Lithium............	Li	6.94	Ytterbium (Neoytter-		
Lutecium...........	Lu	175.0	bium)............	Yb	173.5
Magnesium.........	Mg	24.32	Yttrium...........	Yt	88.7
Manganese.........	Mn	54.93	Zinc..............	Zn	65.37
Mercury...........	Hg	200.6	Zirconium..........	Zr	90.6
Molybdenum.......	Mo	96.0			

LIMESTONE CRUSHER BUILDING AT SYRACUSE QUARRIES OF
THE SOLVAY PROCESS CO.

MODERN LIMESTONE QUARRYING AT THE SYRACUSE QUARRIES.
2,000 POUNDS OF DYNAMITE WERE USED IN THIS SHOT.

TABLE XVI—LINEAR EQUIVALENTS

Mm.	Cm.	Meter	Inch	Foot	Yard	Mile
1.	.1	.001	.03937	.00328	.00109	. . . . . .
10.	1.	.01	.3937	.03280	.01093	. . . . . .
1,000.	100.	1.	39.3700	3.28083	1.09361	.0005681
25.4001	2.5400	.02540	1.	.08333	.02778	. . . . . .
304.801	30.4801	.304801	12.	1.	.33333	. . . . . .
914.402	91.4402	.914402	36.	3.	1.	. . . . . .
1,609,344.	160,934.	1,609.3	63,360.	5,280.	1,760.	1.
						.62137

TABLE XVII—VOLUME EQUIVALENTS

Cu. Meter	Litre	Cu. Cm	Cu. Yd.	Cu. Ft.	Cu. In.	Gal.	Lbs. H_2O @ 15° C.
1.	1,000.	1,000,000.	1.308	35.314	61,022.	264.17	2,202.
.001	1.	1,000.	.0013	.03531	61.022	.2642	2.202
.000001	.001	1.	.0000013	.000035	.06102	.00026	.00022
.7645	764.5	764,500.	1.	27.	46,656.	201.97	1,684.
.02832	28.317	28,317.	.03704	1.	1,728.	7.4805	62.37
.000016	.0164	16.388	.0000215	.00058	1.	.004433	.03608
.003785	3.785	3,785.	.00495	.13368	231.	1.	8.338
.000457	.4569	456.92	.000594	.016038	27.716	1.200	1.

1 liquid oz. = 29.57 cu. centimeters.
1 liquid qt. = 0.9463 liter.
1 liter = 1.0567 liq. qt.
1 dry qt. = 1.101 liters.

1 liter = 0.908 dry qt.
1 peck = 8.81 liters.
1 liter = 1.1135 pecks.
1 bushel = 35.24 liters.

1 hectoliter = 2.8375 bushels.
1 cc. H_2O @ 4° C. weighs 1 g.
1 liter H_2O @ 4° C. weighs 1 kg.
Grams per liter × 58.4 = grams per gal.

TABLE XVIII—SURFACE EQUIVALENTS

Sq. Mm.	Sq. Cm.	Sq. Meter	Sq. Inch	Sq. Foot	Sq. Yd.	Acre
1.	.01	.000001	.00155	.0000108	.00000119	......
100.	1.	.0001	.15500	.00108	......	......
1,000,000.	10,000.	1.	1,550.00	10.7638	1.1960	.000206
645.16	6.4516	.0006645	1.	.00694	.00077	......
92,903.6	929.036	.09290	144.	1.	.11111	......
......	8,361.27	.83613	1,296.	9.	1.	......
......	2,588,881.	......	......	3,097,600.	640
......	4,046.869	......	43,560.	4,840.	1

1 grain T. or A. = 0.0648 grams
1 oz. T. = 31.1035 grams
1 lb. Troy = 0.373 kilograms

TABLE XIX—GRAVIMETRIC EQUIVALENTS

Gram	Kg.	Oz.	Lbs.	Tons
1.	.001	.03527	.0022046	.0000011
1,000.	1.	35.2736	2.2046	.0011023
28.350	.028350	1.	.0625	.00003125
453.60	.45360	16.	1.	.0005
907,180.	907.18	32,000.	2,000.	1.
......	1,000.	......	2,000.	1.1023 = .9842 gross ton

1 gram = 15.432 grains Troy or A.
1 gram = 0.03215 oz. Troy
1 kilogram = 2.679 lbs. Troy

TABLE XX
Comparison of Hydrometer Scales with Baumé Degrees as the Basis

° Baumé	° Twaddell	Spec. Grav.	° Baumé	° Twaddell	Spec. Grav.
1	1.4	1.007	36	66.4	1.332
2	2.8	1.014	37	69.	1.345
3	4.4	1.022	38	71.4	1.357
4	5.8	1.029	39	74.	1.370
5	7.4	1.037	40	76.6	1.383
6	9.	1.045	41	79.4	1.397
7	10.4	1.052	42	82.	1.410
8	12.	1.060	43	84.8	1.424
9	13.4	1.067	44	87.6	1.438
10	15.	1.075	45	90.6	1.453
11	16.6	1.083	46	93.6	1.468
12	18.2	1.091	47	96.6	1.483
13	20.	1.100	48	99.6	1.498
14	21.6	1.108	49	102.8	1.514
15	23.2	1.116	50	106.	1.530
16	25.	1.125	51	109.2	1.546
17	26.8	1.134	52	112.6	1.563
18	28.4	1.142	53	116.	1.580
19	30.4	1.152	54	119.4	1.597
20	32.4	1.162	55	123.	1.615
21	34.2	1.171	56	126.8	1.634
22	36.	1.180	57	130.4	1.652
23	38.	1.190	58	134.2	1.671
24	40.	1.200	59	138.2	1.691
25	42.	1.210	60	142.2	1.711
26	44.	1.220	61	146.4	1.732
27	46.2	1.231	62	150.6	1.753
28	48.2	1.241	63	154.8	1.774
29	50.4	1.252	64	159.2	1.796
30	52.6	1.263	65	163.8	1.819
31	54.8	1.274	66	168.4	1.842
32	57.	1.285			
33	59.4	1.297			
34	61.6	1.308			
35	64.	1.320			

SEMET-SOLVAY COKE OVENS AT DETROIT, MICH., IN WHICH AMMONIA IS PRODUCED FOR USE IN THE ALKALI MANUFACTURE

CLUB HOUSE AND DORMITORY

Social Work

GENERAL WELFARE WORK. Since 1887, The Solvay Process Company has carried on general Welfare Work. A building is devoted to that purpose, with several teachers in charge. Classes are conducted in sewing, dressmaking, cooking and domestic science, both for young children and older girls and women. In a hall especially designed for the purpose, recreation and entertainment of various sorts are provided both for small children and the older boys and girls. Dancing classes, and social dances are held regularly. Recreation rooms are provided for the older boys, where they may play pool and billiards. Also, a gymnasium is provided large enough for basket ball, and where classes are held both for little boys and girls, and for those older. A feature of special importance is the Day Nursery which has proven to be of great use to the mothers of the vicinity. All of this work, of course, is maintained under competent supervision. We wish to emphasize that this department is not limited to employees, but its benefits are open to all residents of the community. The welfare worker who is in charge of the above work also keeps in touch with the employees of the company, and investigates cases of want and appeals for aid.

EDUCATIONAL WORK. The Company conducts within its own works a modern half-time mechanics' school, into which a limited number of boys over 16 years of age are admitted. These boys are paid an hourly wage, and are divided into two classes, each class being alternately two weeks in the school and two weeks in the works. Some very excellent workmen have been trained

GUILD HALL AND GUILD HOUSE

in this school, some of whom are now employed by the company, and some elsewhere.

In addition to the above, in special cases, a plan is provided for loaning money to students to assist them in obtaining a college education.

MUTUAL BENEFIT SOCIETY. In 1888, the Solvay Mutual Benefit Society was organized. This society is really an accident and sickness insurance company. The men and the company pay into the treasury the same amount.

The society employs its own doctor who treats all cases of sickness or accident. Weekly indemnities are paid for a considerable period, in cases of sickness, and accidents occurring off duty. Accidents which occur while on duty are taken care of under the New York State Workmen's Compensation Act.

PENSIONS. Since 1908 the Company has had in operation a general pension plan, to the benefits of which, by action of the Board of Directors, are admitted such of its men as have been incapacitated by reason of old age, sickness or accident after long service.

The amount of pension payable is figured in accordance with carefully considered rules, but is chiefly dependent upon the length of service and the amount received by the beneficiary during the ten highest paid years of his service.

The fund from which pension payments are made was originally set aside out of profits and placed in a separate account. Thereafter, month by month, this fund has been increased by making payments to it of a certain number of cents per ton of product, and all income from the fund has been credited to, and all expenditures and pensions charged against the fund.

The above is a very brief general statement of a part of what The Solvay Process Company has done and is trying to do, along the lines mentioned.

Technical Service Department

REALIZING that in the exceedingly varied use of soda products successful results depend entirely upon the correct chemical application of the product to the process in which it is to be used, this Company has for some time maintained a Technical Service Department composed of a staff of experts devoted solely to the study of the proper use of alkali in all the processes into which it enters.

Even an expert chemist, though he may have very complete knowledge of the chemical properties of a product and of its uses, may not, however, be competent to furnish the most valuable advice and assistance to users of that product. He must, in addition to his chemical knowledge, have carefully studied the processes of manufacture into which the product enters, and he must be thoroughly familiar with the problems and practical difficulties to be encountered in the particular process in which he is seeking to give advice. In other words, he must be a practical operator as well as a chemist.

The experts in the Technical Service Department of The Solvay Process Company are skilled operators in the various industries in which the Company's products are used, and are competent to solve the problems encountered in any process of manufacture into which alkali enters.

This department is maintained for, and is entirely at the service of, consumers of the Company's products. Correspondence addressed to The Solvay Process Company, Syracuse, N. Y., attention Technical Service Department, will receive prompt attention, advice on any problem involving the use of alkali will be gladly furnished, and, where necessary, one of the Company's experts will be sent to study the particular operation and give personal advice and assistance.

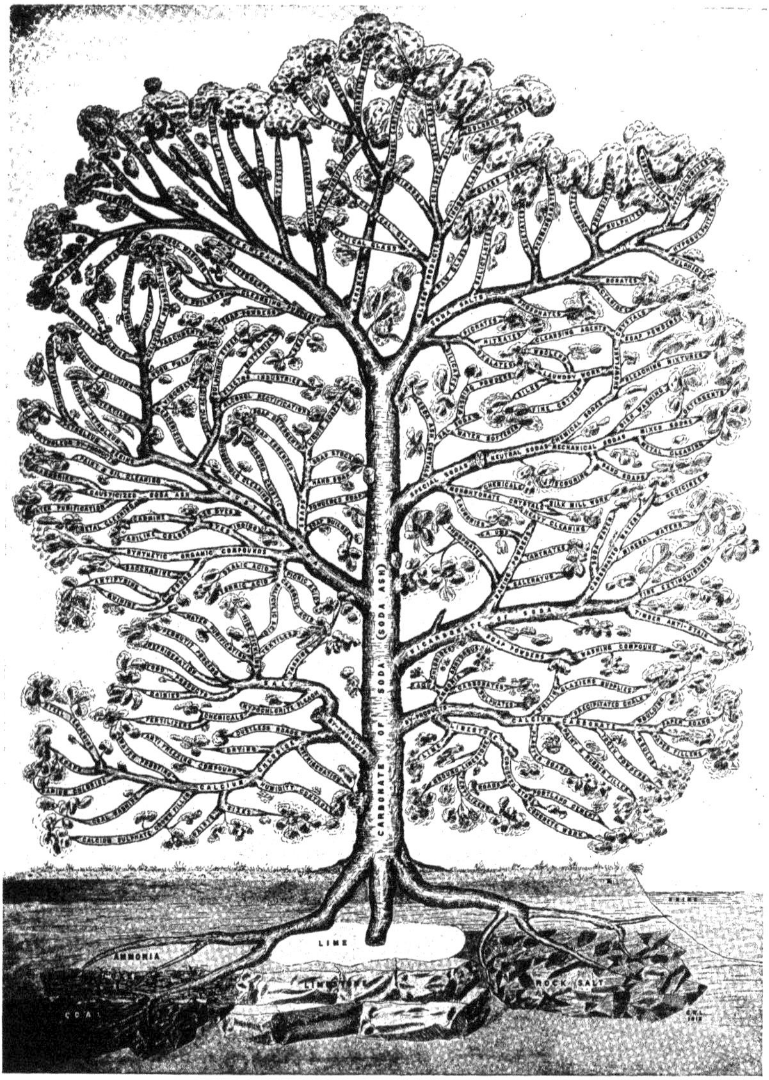

ALKALI TREE

General Index

Index of Tables